82-2528

GOVERNMENT CREDIT SUBSIDIES FOR ENERGY

GOVERNMENT CREDIT SUBSIDIES FOR ENERGY DEVELOPMENT

Murray L. Weidenbaum
Reno Harnish
With James McGowen

American Enterprise Institute for Public Policy Research
Washington, D.C.

Murray L. Weidenbaum is director of the Center for the Study of
American Business at Washington University, St. Louis, Missouri,
and an adjunct scholar at the American Enterprise Institute. Reno
Harnish is a foreign service officer with the State Department. James
McGowen is a graduate student at Washington University.

ISBN 0-8447-3234-6

Library of Congress Catalog Card No. 76-56232

(Energy Policy 2) (AEI Studies 137)

Printed in the United States of America

CONTENTS

PREFACE

This report on proposed credit subsidies for energy development is a joint product. The senior author is Murray L. Weidenbaum, director of the Center for the Study of American Business at Washington University in St. Louis and adjunct scholar at the American Enterprise Institute. Dr. Weidenbaum had general responsibility for the study and wrote Chapters 2, 4, and 5.

Reno Harnish, who was a staff member of the American Enterprise Institute at the time the work was done, wrote the first draft of Chapters 1 and 3. James McGowen, a graduate student at Washington University, prepared the materials on the Reconstruction Finance Corporation and participated in combining the separate contributions into a single report. Emma Williams typed the various drafts of the study.

1

AN ECONOMIC ANALYSIS OF ENERGY SUBSIDIES

Since the embargo imposed in the fall of 1973 by the major foreign oil exporters, the United States has realized the importance of assuring adequate supplies of energy. In addition, the United States has been confronted with the problems that can arise from the inability to meet expected future demands with current domestic sources of supply. The combination of rising demand and declining domestic production has led to a rapid growth in U.S. petroleum imports—from 1.8 million barrels a day in 1960 (19 percent of consumption) to 3.4 million in 1970 (23 percent of consumption) and 6.0 million in 1975 (37 percent of consumption).[1]

A number of far-ranging proposals have been suggested to deal with the situation. They include rationing existing supplies of energy, stockpiling an emergency petroleum reserve, subsidizing conservation and development efforts, relying on the market mechanism to encourage supply and to dampen demand, and declaring a definite point in time as a national goal for achieving energy independence. This report focuses on one particular approach to attaining this goal—the use of governmental credit assistance to encourage the expansion of energy supply in the private sector of the American economy.

Introduction

One of the most ambitious and expensive government programs proposed in recent years was an Energy Independence Authority, largely attributed to Vice President Nelson A. Rockefeller. It comprised a

---

[1] U.S. Federal Energy Administration, *National Energy Outlook: 1976 Executive Summary* (Washington, D.C.: U.S. Government Printing Office, 1976), p. 3.

major part of President Gerald Ford's energy program. The proposal consisted of establishing a government corporation to provide $100 billion to the private sector for developing new domestic energy sources. A more modest version, a $6 billion program of federal loan guarantees to promote a domestic synthetic-fuels industry, is contained in the federal budget for the fiscal year 1977. That proposal was presented as an interim measure, pending establishment of the Energy Independence Authority.

Legislation to establish such an agency, the Energy Independence Authority Act of 1975, was introduced in the 94th Congress. Although the bill was not enacted, similar proposals may be expected in the future.[2] The bill would have established a new government corporation authorized to provide financing and economic assistance for sectors of the national economy which are important to the development of domestic sources of energy or to the conservation of energy. The bill set a target for U.S. achievement of domestic energy independence by 1985.

The corporation—the Energy Independence Authority—would have been authorized to provide a total of $100 billion in loans, loan guarantees, price guarantees, purchases of securities, and other financial assistance to companies that the authority believed would make a significant contribution to increasing the domestic supply of energy through technological and related innovation. The bill made two assertions that provided the rationale for the program: (1) that energy independence and long-term security of energy sources and supplies are "essential" to the health of the national economy, the well-being of our citizens, and the maintenance of national security, and (2) that these goals cannot be attained by relying on "traditional" capital sources in the "traditional" manner.

Several important issues were raised by the Energy Independence Authority proposal. The first is the most fundamental: Why does the domestic energy industry need federal financial assistance in the first place? An immediate corollary is: Are there alternative actions that the federal government could take that would reduce or eliminate the need for new and very expensive programs such as this one?

In addition, other questions arise: How effective are federal credit programs? What are their impacts? How well conceived was this specific program? What would have been the expected results?

This report tries to answer these questions. The remainder of this chapter attempts to shed light on the basic issue of the need for

---

[2] *S. 2532, The Energy Independence Authority Act of 1975*, introduced (by request) by Senators Paul Fannin, Hugh Scott, and John Tower, October 20, 1975.

government subsidy of energy development. Chapter 2 presents an overview of federal credit programs. Chapter 3 examines in detail the economic justification for the first installment of the EIA proposal, the $6 billion synthetic-fuel plan. Chapter 4 contains an examination of the EIA. Chapter 5 summarizes this report, presenting the findings and conclusions.

The Question of Resource Exhaustion. Proponents both of loan guarantees and of other subsidies to potential producers of substitutes for fossil fuels seek to accelerate the development and production of these energy sources, thus preventing an energy shortage when the more common fuels are exhausted.[3] Congressman Mike McCormack (Democrat, Washington) has summarized this position in the following statement: "The central objective of all this is to keep the country from economic and social collapse. If our country is going to survive, it is going to have to have fantastic amounts of new energy . . . the object is to get the energy. And whether it's public or private, we have to get it." [4]

The desirability of the proposed subsidies depends to a great extent on the answers to two questions: How quickly will conventional sources of energy be exhausted? Can automatic market forces operate to stimulate production of substitutes in sufficient quantities before the supplies of fossil fuels are exhausted? If the market can work efficiently without government interference, then the case for government subsidization is substantially weakened. As will be shown, there are strong reasons to believe that the feared hiatus between the exhaustion of conventional fuels and the commercialization of more exotic fuels will not occur.

Professor W. N. Peach of the University of Oklahoma reminds us that a century and a quarter ago the world was worried about running out of trees, then used for both construction and fuel. Coal production at the time was low. But coal quickly became the major source of inanimate energy for most of the world until about the middle of the twentieth century.[5] During that period, there were also times when engineers and geologists lamented the catastrophes that

---

[3] Edward J. Mitchell, *U.S. Energy Policy: A Primer* (Washington, D.C.: American Enterprise Institute, 1974).

[4] Cited in Richard Corrigan, "Energy Report/Rockefeller presses his plan for $100 billion bank for fuels," *National Journal*, October 25, 1975, p. 1472.

[5] W. N. Peach, *The Energy Outlook for the 1980's*, Joint Economic Committee Print (Washington, D.C.: U.S. Government Printing Office, 1973), p. 9.

would follow the imminent exhaustion of coal.[6] A spate of newer articles on the subject [7] includes some empirical studies which apply the theory to the current situation.[8] Historically, however, the notion of absolute resource exhaustion is difficult to support. The typical pattern has not been to "run out" of any specific resource, but for market forces to shift demand to substitutes. The original resource continues to be available, typically at a lower level and higher price.

Historical Experience

The first American switch from exhaustible resources to synthetic fuels and back to exhaustible resources occurred in the nineteenth century. In 1800, illumination in America was provided mainly by oil lamps and candles; the fuel for the lamps was whale oil or sperm oil, both derived from whales. Whale oil and sperm oil are exhaustible resources which differ from fossil fuels in that they are reproducible. Their gradual exhaustion, in the first half of the nineteenth century, caused prices to soar. Sperm oil rose from $.43 a gallon in 1823 to $2.55 a gallon in 1866; whale oil from a low of $.23 in 1832 to $1.45 a gallon in 1865.[9]

As prices increased, consumers switched to cheaper substitutes. The first use of a synthetic fuel, in 1816, was for street lighting, provided by gas manufactured from coal. In the 1830s, cost-cutting innovations in illumination spread to the home. Camphene distilled from vegetable oils rapidly created a new market for lamp light. Another entrant into the low-price, low-quality illumination market was lard oil. In the 1840s, the much-expanded coal-gas industry began to provide illumination for wealthier homes. For several decades, all these fuels continued to compete. But in the 1850s, the

---

[6] C. Robinson, "The Energy 'Crisis' and British Coal," *Hobart Paper No. 59* (London: Institute of Economic Affairs, 1974).

[7] Richard L. Gordon, "A Reinterpretation of the Pure Theory of Exhaustion," *Journal of Political Economy*, June 1967; Robert M. Solow, "The Economics of Resources or the Resources of Economics," *American Economic Review*, May 1974, pp. 1-14; Richard J. Zeckhauser and Milton C. Weinstein, "Optimal Consumption of Depletable Resources," *Quarterly Journal of Economics*, August 1975.

[8] A. S. Manne, "Waiting for the Breeder," *Review of Economic Studies*, 1974; William D. Nordhaus, "The Allocation of Energy Resources," *Brookings Papers on Economic Activity*, no. 3 (1973), pp. 529-76.

[9] Phillip W. Graham, "The Energy Crisis in Perspective," *Wall Street Journal*, November 30, 1973, p. 13.

4

use of coal oil (kerosene) for illumination expanded rapidly because of cost advantages over rival types of energy and quickly came to dominate the residential market.[10]

The success of coal oil was followed by its equally meteoric decline. This abrupt shift occurred as the result of competition from a new fuel that had appeared on the market: crude oil (petroleum), discovered in 1859. As crude-oil production swelled, with few markets for consumption, its price plunged. In January 1860 crude oil sold for $18–$20 a barrel. By the end of 1861 crude oil at the wellhead went for $.10 a barrel. The new low price induced coal-oil refineries to shift to crude as a raw material for kerosene production. By 1863, virtually all coal-oil refineries had shifted to crude-oil refining, and much new refining capacity appeared.

The technological innovations leading to the development of an industry in camphene, lard oil, coal gas, coal oil, and eventually crude oil resulted from an increase in the rising price of the marginal sources of illumination, whale oil and sperm oil. Substitution took time because of the considerable variation in quality of the illuminants which appeared as whale-oil prices rose.

Thus, the theory of exhaustion does not appear to describe this case of resource depletion. Output fell in the whaling industry simply because whale oil had become less profitable. Contrary to the "resource exhaustion" theory, the whale supply was by no means exhausted before consumption shifted to substitute fuels. There was a resource depletion of whales, but it did not proceed in an absolute sense. Whale kill actually increased from the seventeenth century to the mid-nineteenth century. Accessible whaling grounds were exhausted, one after another, but in each case larger and more productive whaling grounds were discovered.[11]

In a period of relatively narrow technical knowledge compared with the present, new illuminants such as coal gas, camphene, lard oil, coal oil, and eventually crude oil were substituted for the old illuminants. Major industries came into being within five years as the price of the marginal source continued to rise. With the introduction of crude oil, illuminant prices fell precipitously. Prior to the exhaustion of whale oil, substitution was complete, and whale oil was no longer competitive. Then, as demand fell, so did the price of whale oil.

---

[10] Harold F. Williamson and Arnold R. Daum, *The American Petroleum Industry: The Age of Illumination, 1859-1899* (Chicago: Northwestern University Press, 1959), p. 13.

[11] Albert Cook Church, *Whale Ships and Whaling* (New York: W. W. Norton & Co., 1960).

The Current Situation

Let us now turn to the current situation. Is the United States approaching exhaustion of its fossil fuels such as crude oil and natural gas? Would the economy produce synthetic-fuel substitutes without loan guarantees? When would synthetic fuels be developed?

The government's price policy of the last twenty years toward both petroleum and natural gas has been to prevent the nation's producers from concentrating on the substitution of synthetic fuels. Through the price control programs, government policy is bringing about—on at least a temporary basis and perhaps unwittingly—what it is seeking to prevent through the proposed synthetic-fuels programs.

Other factors are also at work to delay the development of synthetic fuels. In the last five years, estimates of the commercial selling price of synthetic crude oil have risen sharply. In 1970, the price of oil shale, including a 12 percent rate of return and a resource charge for shale (royalty) of $.40, was estimated at between $3.60 and $4.35 a barrel.[12] In 1973, stated in 1970 prices with a 10 percent return and no royalty, the estimated price was $5.58 a barrel for high-grade shale; in 1974 the projected cost in 1973 prices with a 15 percent rate of return was $6.80 a barrel; in 1975 estimates ran in the neighborhood of $15 a barrel.[13] In five years the estimated price of crude oil from shale thus increased 310 percent.

The picture is similar for price estimates of gas derived from coal. In 1971, a price of $.33 per thousand cubic feet was reported; in 1973, William D. Nordhaus projected synthetic gas at $1.19 per thousand cubic feet; and by 1975 the President's task force on synthetic fuel reported a cost of approximately $2.70 per thousand.[14] In four years these rough cost estimates had risen 710 percent. Such a path for projected costs tends to give pause to the potential investor.

Other price uncertainties are caused by the possibility of continued controls on crude-oil and natural-gas prices. Since prices for exhaustible resources on the U.S. market in 1985 are likely to be lower than the prices of synthetic fuels, when would synthetics come into production in the absence of government assistance? One source calculates that, given their cost relative to other fuels, synthetic oil

---

[12] Cabinet Task Force on Oil Import Control, *The Oil Import Question* (Washington, D.C.: U.S. Government Printing Office, 1970), p. 305.

[13] Nordhaus, "Allocation of Energy," p. 544; M.I.T. Energy Laboratory Policy Study Group, *Energy Self-Sufficiency* (Washington, D.C.: American Enterprise Institute, 1974), pp. 52-53.

[14] Nordhaus, "Allocation of Energy," p. 544.

and gas would optimally come into the market sometime between the years 2010 and 2020.[15]

Some Policy Implications. The price increase associated with the exhaustion of a resource can produce a smooth transition to substitute forms and simultaneously open up commercial opportunities for technological innovation. This conclusion should be even more relevant in the current age of rapid technological change than it was in the first half of the nineteenth century. However, government energy policies of the last twenty years, including quotas and price controls on oil and gas, have interfered with the smooth market adjustment to substitute fuels. Continuation of price controls will increase the size of the domestic shortage, as producers develop expectations of decontrol with a consequent increase in royalties, while consumers are encouraged to use energy at prices below the true economic cost. Under such conditions, higher-cost synthetic production without government assistance appears to be a low probability for the coming decade.

Subsequent chapters examine the case for federal financial assistance as a means of encouraging expansion of domestic energy production. This study does not deal with financing the exploratory research and development work that may give rise to successive technological innovations, an area where government support is a long standing tradition. This support is justified by the "externalities" produced—the benefits that are not necessarily captured by the individual or organization doing the work, but that are available to the entire society.

In the energy field, public policy is quite clear. Large and rising amounts of public funds are being devoted to energy development. The federal budget for the fiscal year 1977 calls for appropriations of $6.0 billion for the Energy Research and Development Administration, an increase of more than 70 percent from the 1975 level of $3.5 billion.

In striking contrast, deliberate government efforts to stimulate or "force-feed" the commercialization of new technology have been in the main disappointing. Advances have been made, to be sure. Important new commercial products have come from or been assisted by government-supported R & D. Jet engines, computers, and microminiaturization are all cases in point. But in each of these successful instances, the movement from the laboratory to commercial produc-

---

[15] Nordhaus, "Allocation of Energy," p. 552; J. A. Hausman, "Project Independence Report: An Appraisal of U.S. Energy Needs Up to 1985," *Bell Journal of Economics*, Autumn 1975, p. 548.

tion and sales resulted from the operation of market forces, and not from deliberate government efforts to foster commercial products.

Over the past thirty years, deliberate efforts to transfer the products of government-sponsored R & D to the private sector have resulted mainly in unsuccessful projects which eventually were abandoned—ranging from hoped-for breakthroughs in housing to new consumer products from aerospace companies. The outcome was a waste of taxpayers' money and a weakening of the companies' capability to finance their traditional activities.[16]

---

[16] See M. L. Weidenbaum, *The Economics of Peacetime Defense* (New York: Praeger Publishers, 1974), especially "Transferability of Defense Technology," pp. 140-44; M. L. Weidenbaum, *The Modern Public Sector* (New York: Basic Books, 1969), pp. 67-75.

2

IMPACTS OF GOVERNMENT CREDIT PROGRAMS

Any evaluation of proposals for governmental credit assistance in the energy area should be made in the context of the substantial experience with the wide array of existing federal credit activity. Since most of the credit programs do not appear in the federal budget, they seem to be a painless way of achieving national objectives. According to their proponents, the federal government is "merely" guaranteeing private borrowing or sponsoring ostensibly private institutions, albeit with federal aid. Existing examples include the federal land banks and the federal home loan banks. The proposed Energy Independence Authority would constitute a major expansion of government credit programs.

Is this use of the federal government's credit power a variation of the proverbial "free lunch"? As will be demonstrated, use of the governmental credit power does result in substantial costs to business as well as to taxpayers, but it generates opportunities for the application of federal controls over private economic activity, making the added regulation more palatable. On the other hand, substantial benefits may accrue from these programs in achieving various national priorities. The advantages of government credit power arise from channeling credit—and ultimately additional real resources—to specific groups of the society. In each government credit program, Congress has passed a law stating in effect that the national welfare requires designated groups to receive larger shares of the available supply of credit than would result from the operation of market forces alone.

Less apparent than the benefits are the costs and other side effects of the expanded use of government credit programs. In terms of overall economic impact, the programs do little to increase the total pool of capital available to the economy. They result in a game of musical chairs. By preempting a major portion of the annual flow of savings, the government-sponsored credit agencies reduce the amount of credit that can be provided to unprotected borrowers, mainly consumers, state and local governments, and private business firms.

During periods of tight money, it is difficult for unassisted borrowers to attract the financing that they require. They are forced to compete against the government-aided borrowers. Federal loan guarantees reduce the risk of lending money to the insured borrowers. The uneven competition results in still higher interest rates. The supporters of the proposed government credit subsidies for energy development have admitted to this effect. The following is taken from the congressional testimony of Gerald L. Parsky, assistant secretary of the Treasury, on proposed federal financial incentives for synthetic-fuel demonstration plants: "Such incentives increase the demand for capital while having little or no effect on the overall supply of capital. They tend to cause interest rates to rise and channel capital away from more economic to less economic uses." [1]

A sympathetic investment banker testifying on the same program offered a complementary analysis:

> The proposed $6 billion government guarantee program, however, should not be taken too lightly, especially under the assumption that additional guarantee programs for other energy sources will also be implemented. The ability of smaller companies and companies with lower credit ratings to compete for available capital could be lessened at times. In addition, higher grade corporate debt securities could face slightly higher interest rates in order to compete with an increased supply of government guaranteed debt. [2]

Impacts on Total Saving and Investment

The conclusions of the professional literature on the impact of federal credit programs on the total flow of saving and investment in the American economy are clear. These programs do little or nothing to increase the total flow of saving and investment. Instead, they change the share of investment funds going to a given industry or

---

[1] Gerald L. Parsky, Assistant Secretary of the Treasury, Statement before the House Committee on Science and Technology, Subcommittee on Energy Research, Development, and Demonstration (Fossil Fuels), October 22, 1975, p. 6.

[2] Arthur B. Treman, Jr., Statement before the House Committee on Science and Technology, October 20, 1975, p. 13.

sector of the economy and, in the process, exert upward pressures on interest rates as investment funds are attracted away from other sectors.

In commenting on existing programs of federally assisted credit to the private sector, Dr. Henry Kaufman, economist with the investment house of Salomon Brothers, has written: "Federal agency financing does not do anything directly to enlarge the supply of saving. . . . In contrast, as agency financing bids for the limited supply of savings with other credit demanders, it helps to bid up the price of money." [3]

In referring to borrowing by the federal government and its agencies, Dr. Albert Wojnilower has made a similar observation: "Because these governmental borrowers need have few if any worries about creditworthiness or meeting interest payments, they can preempt as much of the credit markets as they choose. As a result, the Federal sector has become one of the most relentless sources of upward pressures on interest rates." [4]

In a comprehensive study of federal credit programs for the prestigious Commission on Money and Credit, Warren Law of Harvard University concluded that the programs have created inflationary pressures in every year since World War II.[5] Professor Patricia Bowers has noted what she terms the "costs" of federal credit programs. For example, given the availability of funds, an increase in credit in the area of housing means less credit for other borrowers, chiefly state and local governments and small businesses. Not only is less money available, but the operations of the federal credit agencies tend to increase interest rates above the levels that would have prevailed if they had not entered the credit markets.[6]

A variety of factors are responsible for the resulting tight credit and high interest rates. The total supply of funds is broadly determined by household and business saving and by the ability of banks

---

[3] Henry Kaufman, "Federal Debt Management: An Economist's View from the Marketplace," in Federal Reserve Bank of Boston, *Issues in Federal Debt Management*, 1973, p. 171.

[4] Albert M. Wojnilower, "Can Capital-Market Controls Be Avoided in the 1970's," in Michael E. Levy, ed., *Containing Inflation in the Environment of the 1970's* (New York: Conference Board, 1971), p. 42.

[5] Warren A. Law, "The Aggregate Impact of Federal Credit Programs on the Economy," in Commission on Money and Credit, *Federal Credit Programs* (Englewood Cliffs, N.J.: Prentice-Hall, 1963), p. 310.

[6] Patricia F. Bowers, *Private Choice and Public Welfare* (Hinsdale, Ill.: Dryden Press, 1974), pp. 494-96. See also Alan Greenspan, "A General View of Inflation in the United States," in Conference Board, *Inflation in the United States* (New York: The Board, 1974), p. 4.

11

to increase the money supply. These conditions set the basic limit on the availability of funds referred to by Professor Bowers. The normal response of financial markets to an increase in the demand for funds by a borrower, such as a federal credit program, is an increase in interest rates to balance the demand for funds with the supply of saving. But the federal government's demands for funds are "interest-inelastic" (the Treasury will generally raise the money that it requires regardless of the interest rate), and the interest-elasticity of saving is relatively modest. Thus, weak and marginal borrowers will be "rationed out" of financial markets in the process, while the Treasury and other borrowers pay higher rates of interest.

The General Accounting Office has noted these impacts in its analysis of the proposed Energy Independence Authority:

> In this connection we note that the bill is not neutral on conservation options. Actually, it would hamper conservation efforts rather than simply fail to promote them. This is true because the bill would result in allocation, not creation, of capital. The EIA's loan funds would, in large part, be raised in the private capital market. Its guarantees would make projects it assists financially more attractive to private capital than conservation projects not backed by Federal guarantees. Thus, both its loans and its guarantees will siphon private capital away from those conservation projects which might have been able to obtain private financing in the absence of EIA operations.[7]

Important insight into the effects of federal credit programs on capital markets has been provided by Bruce MacLaury, the president of the Federal Reserve Bank of Minneapolis and a former deputy undersecretary of the Treasury: "The more or less unfettered expansion of Federal credit programs . . . has undoubtedly permitted Congress and the Administration to claim that wonder of wonders—something for nothing, or almost nothing. But as with all such sleight-of-hand feats, the truth is somewhat different." [8]

There are extra costs associated with introducing new government credit agencies to the capital markets and selling securities that only approximate the characteristics of direct government debt. As a result of such considerations, the market normally charges a premium over the interest cost on direct government debt of comparable

---

[7] Comptroller General of the U.S., *Comments on the Administration's Proposed Synthetic Fuels Commercialization Program* (Washington, D.C.: General Accounting Office, 1976), p. 23.

[8] Bruce K. MacLaury, "Federal Credit Programs—the Issues They Raise," in Federal Reserve Bank of Boston, *Issues in Federal Debt Management*, 1973, p. 214.

maturity. That premium ranges from 0.25 percent on the well-known federally sponsored agencies such as Federal National Mortgage Association to more than 0.5 percent on such exotics as New Community Bonds. In general, if cost of financing were the only consideration, it would be more efficient to have the Treasury itself provide the financing for direct loans by issuing government debt in the market.

Reduced efficiency occurs in the economy by providing a federal "umbrella" over credit activities without distinguishing their relative credit risks. A basic function that credit markets are supposed to perform is that of distinguishing credit risks and assigning appropriate risk premiums. This function is the essence of the ultimate resource allocation of credit markets. As an increasing proportion of issues coming to the credit markets bears the guarantee of the federal government, the ability of the market to differentiate credit risks inevitably diminishes. Theoretically, the federal agencies issuing or guaranteeing debt perform this role, charging as costs of the programs differing rates of insurance premiums. In practice, all of the pressures are against such differential pricing of risks.[9] The abandonment of this function is a hidden cost of federal regulation via credit programs.

Professor Henry D. Jacoby, of the Massachusetts Institute of Technology, discussed the costs of government credit programs while advocating a limited program of loan guarantees for synthetic-fuel development: "The problem with loan guarantees is that they tend to hide the true cost of the technology that is being demonstrated. . . . Thus the guarantee carries a hidden subsidy which masks the real economic cost of the energy produced—or saved—and clouds the issue of what the 'commercial' status of the technology would be without the guarantee."[10]

Impacts on Sectors of the Economy

The nature of federal credit assistance is to create advantages for some borrowers and disadvantages for others. The groups which tend to be eliminated in the process are unlikely to be the large well-known corporations or the U.S. government. They are more likely to be state and local governments, medium-sized businesses, private mortgage borrowers not under the federal umbrella, and consumers,

---

[9] Ibid., p. 217.

[10] Dr. Henry D. Jacoby, Massachusetts Institute of Technology, Statement before House Committee on Science and Technology, *Synthetic Fuel Loan Guarantees,* 94th Cong., 2d sess., 1976, vol. 1, p. 370.

Table 1
IMPACT ON CREDIT MARKETS OF FEDERAL AND FEDERALLY ASSISTED BORROWING
(fiscal years, dollars in billions)

Category of Credit	1960	1965	1970	1975
A. Federal borrowing	$ 2.2	$ 4.0	$ 5.4	$ 50.9
B. Federally assisted borrowing (off-budget)[a]	$ 3.3	$ 6.8	$15.1	$ 13.9
C. Total federal and federally assisted borrowing (A + B)	$ 5.5	$10.8	$20.5	$ 64.8
D. Total funds advanced in credit markets	$43.4	$69.6	$89.0	$177.9
E. Federal percentage of total[b]	12.7%	15.5%	23.0%	36.4%

[a] Obligations issued by government-sponsored agencies or guaranteed by federal agencies.
[b] (C)/(D).
Source: Federal Reserve System; U.S. Department of the Treasury.

thereby contributing to the economic and financial concentration in the United States.

The competition for funds by the rapidly expanding federal credit programs also increases the cost to the taxpayer by raising the interest rate at which the Treasury borrows its own funds. As shown in Table 1, there has been a massive expansion in the size and relative importance of federal government credit demands over the past decade. In 1960, the federal share of funds raised in private capital markets, using the Federal Reserve System's flow-of-funds data, was 12.7 percent. The government's share rose to 23 percent in 1970 and to over 36 percent in 1975.

Virtually every session of Congress in recent years has enacted additional federal credit programs. Since 1960, the Federal National Mortgage Association (Fannie Mae) has been joined by the Government National Mortgage Association (Ginnie Mae), Student Loan Marketing Association (Sally Mae), and the U.S. Railway Association.

Relation to Government Controls. An examination of existing programs of federal guarantee of private credit reveals that credit assistance is often accompanied by various forms of governmental controls or influence over its recipients. For example, federal credit guarantees for shipbuilders are accompanied by requirements that various "national defense" features be incorporated into the vessels. The largest federal program for guaranteeing private credit, that administered by the Federal Housing Administration (FHA), demonstrates the extent to which controls may accompany the credit assistance. The FHA conducts an inspection of each residence to determine whether the builder has abided by all of the agency's rules and regulations governing the construction of the homes that it ensures. There are four separate "veto" points facing a builder applying for FHA insurance of mortgages for a new project: (1) affirmative marketing to minority groups, (2) environmental impact, (3) architectural review, and (4) underwriting.

The division of responsibilities among the various federal housing offices may cause considerable confusion and delay. For example, after the underwriting has been approved, giving an appraised value high enough to cover the builder's cost, additional requirements may be imposed by the environmental impact office or by the architectural review, substantially raising the cost of the project. If this occurs, the builder must return to the appraiser's office and attempt to obtain a revised underwriting.

Miles Colean, analyst of the housing industry, has commented on the deleterious effects of the increasing array of government controls imposed via the FHA program: "The complications of FHA operations, by introducing numerous requirements irrelevant to the extension of mortgage credit, placed the market oriented activity of FHA at a competitive disadvantage." [11] Despite its subsidy element, the FHA program has been a declining factor in the housing market in recent years, mainly for the reasons described above.

The RFC Experience. The most ambitious governmental effort to provide credit to private business to date has been the Reconstruction Finance Corporation (RFC). Under the original act passed in 1932, Congress granted the RFC lending powers which extended only to railroads and financial institutions. By 1934, Congress had come to believe that the industrial sector also required the resources of the new federal corporation. Continuing economic problems led to a further expansion of the RFC's powers during the late 1930s. In 1938, Congress enacted a law which gave the RFC the authority to purchase the securities and obligations of any business enterprise.[12]

---

[11] Miles L. Colean, "Quarterly Economic Report," *Mortgage Banker*, March 1974, p. 63.

[12] Reconstruction Finance Corporation Act of 1932, Public Law 2, 72d Cong., January 22, 1932; Public Law 417, 73rd Cong., June 19, 1934; Public Law 1, 74th Cong., January 31, 1935; Public Law 479, 75th Cong., April 13, 1938.

In the late 1940s the criteria used by the RFC to approve loans were extremely vague. The practical guide of profitability was not applicable to its loans since the corporation was prohibited from competing with private sources of credit. Hence, a large portion of the loan applications to the RFC were of the type considered to be poor risks by private credit institutions.

Congress stipulated that the RFC should extend loans only for purposes which would serve the public interest. The RFC, however, apparently paid little attention to this criterion. The test of public interest, for example, would have called for a reduction in the volume of loans during the period of inflationary pressures following 1948. Instead, the corporation actually expanded its loan activities during this period. Many loans were difficult to justify in terms of the formal criterion.[13] Table 2 provides examples of loans to gambling casinos, bars, breweries, and similar establishments.

In 1949, rumors began to circulate that often connections with influential people in Washington were the real criterion for gaining loan approval from the RFC. When these reports persisted, the Senate Committee on Banking and Currency formed a subcommittee chaired by Senator J. William Fulbright to investigate the corporation's lending practices.

The Fulbright subcommittee disclosed numerous examples of favoritism and corruption in the granting of RFC loans. Two of the RFC's five directors and one of President Harry S. Truman's closest advisers were charged with influence-peddling activities. Other examples of questionable practices included the RFC loans to the Lustron Corporation and the Kaiser-Frazer Corporation. In the Lustron case, Carl Strandlund, the president and principal owner of Lustron, charged that RFC director Dunham repeatedly tried to transfer control of the corporation to some of his friends and associates. Evidence also indicated that Edgar Kaiser of the Kaiser-Frazer Corporation had been approached by influence peddlers on several occasions.[14]

The history of the RFC suggests several problems that may be associated with government credit programs:

- The criteria for granting federal loans are likely to be vague and open to subjective interpretation.

---

[13] Jules Abels, The Truman Scandals (Chicago: Henry Regnery Co., 1956), pp. 70-82.

[14] Ibid., pp. 10-11; U.S. Congress, Senate, Committee on Banking and Currency, Study of Reconstruction Finance Corporation and Proposed Amendment of RFC Act, Senate Report 649, 82d Cong., 1st sess., 1951, p. 41; U.S. Congress, Senate, Committee on Banking and Currency, Progress Report on RFC Liquidation, Hearings, 83rd Cong., 2d sess., 1951, p. 54.

Table 2

SELECTED RFC BUSINESS LOANS OUTSTANDING, 1950

Loan Recipient	Amount Outstanding
United Distillers of America, New York, N.Y.	$331,500
James Distillery, Inc., Baltimore, Md.	315,000
Harvard Brewing Co., Lowell, Mass.	300,000
Bluebeard's Castle Hotel Corp., St. Thomas, Virgin Islands	250,000
Morello Winery, Kerman, Calif.	200,000
Coast Drive-In Theaters, Calif.	164,669
Shore Club Lodge, Inc., Boise, Ida.	164,500
Saratoga Hotel Co., Saratoga, Wyo.	125,000
Coast View Winery, Inc., Fresno, Calif.	117,750
Gold Front Bar, Gold Front Theatre, Gold Front Recreation, Sheboygan, Mich.	85,000
Wooden Shoe Brewing Co., Minster, Ohio	65,040
Plymouth Rock Bar, Detroit, Mich.	39,500
Sandpiper Inn, Fort Walton, Fla.	32,000
Cactus Courts, Carlsbad, N. Mex.	11,446

Source: Testimony of Herbert C. Hoover in U.S. Congress, Senate, Committee on Banking and Currency, *Hearings on S. 514, S. 1116, S. 1329, S. 1376, and S. J. Res. 44,* 82d Cong., 1st sess., 1951, pp. 91-93.

- Government subsidization of business often encourages and perpetuates a misallocation of resources.
- Government programs tend to develop a life of their own and to persist long after the problems for which they were created have been solved.
- Federal credit programs put the government in the position of holding assets of questionable quality or limited use, making it difficult to recover the original value of loans in the case of default, and complicating the process of liquidating the agency.

Summary

Boiled down to basics, federal credit programs merely shift funds from one potential borrower to another. They do not increase the amount of funds available to the economy. Rather, to the extent they succeed, they take capital away from the unassisted sectors of the economy and lead these sectors to request aid themselves. Government guarantees also tend to raise interest rates for private as well as government borrowers. Quite clearly, federal credit subsidies for energy development would possess all of these limitations and shortcomings.

An alternative to proliferating the already large array of federal credit programs is to deal with the underlying conditions of which they are symptoms. If we can create an economic climate more conducive to private saving and investment, the need for private borrowers to seek federal credit assistance will be reduced.

What can be done to provide greater encouragement to saving and thus to increase the basic supply of investment capital? The first and perhaps most important idea is essentially a negative one. The federal government should stop being such a large dissaver—that is, it should eliminate or at least reduce the massive extent to which it currently competes with the private sector for the relatively limited supply of investment capital. As the economy continues to recover from its recession lows, the rising pace of business activity will yield increasing flows of federal revenues. Unless Congress increases government spending at that same rapid rate, the effect will be a substantial reduction in the federal deficit in the fiscal years 1977 and 1978. This result, however, will not be automatic. The advocates of economy will have to exert sufficient political pressure to offset the proponents of greater government spending.

A related but far more technical approach has little public support or even understanding: the need to curtail the various off-budget agencies. These agencies provide subterfuges to disguise normal federal expenditures by wholly federal agencies so that they do not show up in the budget. Not only do the expenditures continue, but, because they are no longer subject to the scrutiny of the budgetary process, they expand at a far more rapid rate. In fiscal 1972, they totaled $249 million. In the fiscal 1977 budget they are estimated at over $10 billion—money that the U.S. government has to borrow above and beyond the official budget deficit. Should a proposal like that for an off-budget Energy Independence Authority be adopted, off-budget expenditures would more than triple.

A second useful contribution that the federal government can make to ensure capital adequacy in the years ahead lies in the area of government controls over business. An increasing number of regulatory agencies impose investment requirements on business firms, investments that do not generate more productive capacity but that are intended to meet various social priorities. Both public and private projections show that rising annual dollar outlays for new pollution control facilities will be needed to meet existing legal requirements. About 5 percent of industrial plant and equipment investments is expected to be devoted to these purposes. In addition, government-mandated industrial safety and noise abatement outlays will be significant, with estimates ranging to $40 billion or more during the coming five-year period.[15] It is not proposed here that these social requirements be eliminated but, rather, that they be subject to the rigors of a cost-benefit test. These expensive federal regulatory requirements should be continued only where it can be demonstrated that their value or benefit to the society exceeds the costs they impose on the public. As shown in a subsequent chapter, regulatory requirements are especially severe in the energy area.

In addition, techniques are available that would provide positive incentives to saving or investment or both. One economically efficient approach to increasing private saving is to reduce the corporate income tax. That action would have a number of desirable effects. The amount of business "saving" would be increased in the form of retained earnings. The portion of the tax reduction not saved would be disbursed in the form of higher dividends, and individual disposable income and personal saving would rise. The tax saving might also be passed along in two directions: forward to consumers in the form of lower prices, or rather more slowly rising prices, and backward to labor in the form of higher wages, salaries, and fringe benefits.

In addition, a lower corporate income tax rate would reduce the indirect but pervasive role of the tax collector in internal business decision making. It would tend to promote more efficient use of resources to the extent that fewer low-priority business expenses would be incurred merely because they are tax deductible. It would soften the double taxation of corporate income. A lower corporate income tax would also reduce the current bias in the tax system toward debt financing—because interest paid on debt is deductible from taxable income, and in most cases dividends on equity capital are not. Rising debt/equity ratios and declining interest coverages on corporate balance sheets clearly demonstrate the importance of permitting a greater reliance on equity rather than on debt financing in the future.

---

[15] Murray L. Weidenbaum, *Government-Mandated Price Increases* (Washington, D.C.: American Enterprise Institute for Public Policy Research, 1975).

On the other hand, unlike earlier suggestions for reduced government spending and controls, tax cuts would increase the federal deficit and thus the government borrowing that competes with private investment demands. The beneficial impacts on production and employment of a cut in corporate income taxes would generate "feedback" effects and result in significant compensating increases in federal revenues. Encouragement to individual or consumer saving could be accomplished through excluding from gross income all or a portion of interest on savings. However, these actions also would adversely affect efforts to reduce federal budget deficits.

3
SUBSIDIES FOR SYNTHETIC
FUEL DEVELOPMENT

The Ford administration proposed the use of federal loans and loan guarantees to the producers of synthetic fuels as a means of attaining the goal of energy independence. The President emphasized that the program would bring increased national security and an improved performance of the nation's economy: "America cannot permit the excessive delays associated with commercialization of unconventional energy technologies. New production is essential. Our national security and economic well-being depend on our ability to act decisively on energy." [1]

The central questions are whether a federally subsidized acceleration of synthetic-fuels production will actually achieve the intended benefits, and, if so, whether these benefits will outweigh the costs of such a program. These questions have been analyzed by the Synfuels Interagency Task Force of the President's Energy Resources Council, [2] which examined the impact of three alternative synthetic-fuel programs ranging in output size from 350,000 to 1.7 million barrels a day.

The task force found that expected costs exceeded expected benefits at every level of synthetic-fuel production which it considered. The remainder of this chapter will discuss the task force report in detail and will point out additional factors that make the commercialization of synthetic fuels appear unfavorable at the present time.

---

[1] Text of letters from the President to the speaker of the House of Representatives and the president of the Senate, October 10, 1975, Office of the White House Press Secretary.

[2] President's Energy Resources Council, Synfuels Interagency Task Force, *Recommendations for a Synthetic Fuels Commercialization Program*, 3 vols. (Washington, D.C.: U.S. Government Printing Office, 1975).

Table 3

EXPECTED DISCOUNTED NET BENEFIT OF SYNTHETIC FUEL PROGRAM

(in billions of 1975 dollars)

Program Alternative	Economic Benefit	Embargo Protection	Environmental and Socio-economic Cost	Total
No program	0	0	0	0
Information program (0.35 mm bbl/day)	−1.64	0.43	−0.44	−1.65
Medium program (1 mm bbl/day)	−5.45	1.18	−1.14	−5.41
Maximum program (1.7 mm bbl/day)	−11.22	2.23	−1.99	−10.98

Source: Synfuels Task Force, *Recommendations for Synthetic Fuels*, vol. 2, p. 63.

Summary of Program Effects

The estimated cost to the nation of undertaking a commercialization program to produce 350,000 barrels a day of synthetic fuel is $1.6 billion, in 1975 dollars (see Table 3). The expected cost to the nation of undertaking a 1-million-barrels-a-day program is $5.4 billion, and the anticipated cost of the "maximum" program (1.7 million barrels a day) is $11 billion. In these estimates, the costs have been discounted to their present value and therefore represent net changes from the situation which would prevail without the program. The costs and benefits included in this analysis cover market-priced items such as energy and factors of production, as well as estimates of environmental pollution and benefits to national security.

The figures presented are the mean or average values selected from a range of possible costs and benefits. A synthetic-fuels program is affected by many unpredictable factors.[3] As shown in a previous chapter, estimates of synthetic-fuel costs have increased several hundred percent in the last few years, and no one can be certain what they will be in 1985. Neither is it possible to predict the future price strategy of the oil cartel. Indeed, no one knows whether

---

[3] For a critical evaluation of the Synfuels Task Force study by the Library of Congress, see Martha Krebs-Leidecker, "Government Ownership of Synthetic Fuels Plants—a Pro-Con Analysis," in House Committee on Science and Technology, *Synthetic Fuel Loan Guarantees*, vol. 2, pp. 91-99.

the cartel will exist ten years from now, or if there will ever be another oil embargo. To take these uncertainties into account, the Synfuels Interagency Task Force made judgments about the probability of these factors occurring, and then combined the individual probabilities to show various possible outcomes. In the case of the 350,000-barrels-a-day program, there is a 10 percent chance that the country will lose more than $9 billion and a 10 percent chance the country will gain more than $7 billion. However, there is a 70 percent chance that the country will have less output from its employment, capital, and raw materials after the program than if we do not undertake it.

Economic Well-Being. The task force report indicates that a synthetic-fuel subsidy program would lead to a decline in economic welfare because it would shift productive inputs to areas where their output was valued less than previously. The report estimates that the implementation of a synthetic-fuel program, depending upon whether it is established at the level of 350,000 barrels a day or 1.7 million barrels a day, will cause the value of U.S. production in 1985 to fall the equivalent of $1.5 billion to $11 billion.[4]

Field price controls on natural gas currently limit the price which can be paid for new gas to $.50 per thousand cubic feet. Pipelines purchase the gas from producers, transport it to market areas, and then resell it to gas utilities or industrial users. Pipeline tariffs are also regulated. Therefore, the market for gas at the city gate can be represented as a price-controlled market. Several effects are immediately apparent. Since the price is controlled below equilibrium, more gas is demanded than supplied at the market price, resulting in a shortage. More important, gas utilities, which are the buyers in this case, would be willing to offer as much as $3.00 per thousand cubic feet for the next unit of gas, but society is only devoting $.80 per thousand cubic feet worth of resources toward obtaining more gas. Without price controls, resources valued by consumers at $.81 a unit elsewhere could be attracted into the gas industry to produce another unit of output valued by consumers at $3.00. In theory, society would be benefited by moving resources into the industry until the value of the output to consumers was equal to the amount it took to attract the last increment of factors into the industry. But since the resources are not available, gas utilities would be forced into the alternative of synthetic gas which, though also valued at $3.00 for the first unit by consumers, would use up $3.00 worth of

---

[4] Synfuels Interagency Task Force, *Recommendations*, vol. 1, p. 63.

inputs. This government-caused misallocation was recognized in the task force report:

> A possible ironic outcome is that economically inefficient price regulation leads to a shortage of relatively cheap natural gas which is "ameliorated" by the economically inefficient production of expensive synthetic gas. In other words, if deregulation of natural gas were to take place, consumers may benefit in that the quantity of gas supplied would be the same but at a lower aggregate cost than under the synthetic natural gas program.[5]

Because of the nature of utility regulation and the tendency for rates to be based on a target return on investment, gas companies would still tend to build their own synthetic-fuel plants, even though the basic economics might be unfavorable.

National Security. Federal subsidies for the development of synthetic fuels would not contribute to future economic welfare, but would they advance the goal of national security? General Maxwell D. Taylor defined the role of national security as "to protect those things we consider indispensable to our survival, power or well-being and hence deserving the expenditure of effort and resources to gain, retain or enjoy." He went on to say that "one could hardly hope to find a better example of the seriousness of nonmilitary threats to national security than the present energy crisis. Our national valuables in many guises and over a wide span of our interest are endangered."[6]

General Taylor was describing the embargo and price rise of 1973–74. For the present, a discussion of the relation between energy and the national security can be limited to the question of oil imports. Other relationships between energy and national security, such as the need to protect domestic industry, were examined and rejected by the Cabinet Task Force on Oil Import Control in February 1970.[7] The task force listed eight types of national security risks to be avoided, including the possibilities that (1) a group of producers might act in concert to deny oil; (2) exporting countries might take over assets of U.S. companies; and (3) exporting countries might

---

[5] Ibid., p. 68.

[6] Maxwell D. Taylor, "The Legitimate Claims of National Security," *Foreign Affairs*, April 1974.

[7] Cabinet Task Force on Oil Import Control, *The Oil Import Question* (Washington, D.C.: U.S. Government Printing Office, 1970), p. 31.

form an effective cartel to raise prices. In the intervening period, these three national security risks have become actualities.

The Federal Energy Administration's *Project Independence Report* estimated the cost of embargo, production cutback, and price rise at $10–20 billion in terms of lost U.S. gross national product. Further, the FEA found that if world oil prices were $11 a barrel and a one-year embargo occurred in 1985, it might cost the U.S. economy $30–40 billion.[8] By General Taylor's definition, a continuing threat to national security is posed by the forecast for U.S. energy demand and supply in 1985. The threat would seem to justify paying some level of insurance in the present. The challenge is to provide this insurance at the least cost, and the cost of the method depends greatly on the world price of oil.

National security insurance benefits from synthetics must be compared with the cost to the economy of using high-priced energy. The benefit of such programs will be limited by the fact that guaranteed loans for synthetic and shale production mean less credit available for wildcat wells and electric utility plants: this is the "crowding out" problem, discussed in the previous chapter on government credit programs.

If world oil prices fall, there will be a greater number of defaults, and the United States will have paid a larger premium for insurance against the national security risk. In an attempt to quantify the "insurance" value of synthetic fuels, the President's Synfuel Task Force assumed that the chances of an embargo were one in ten for 1985 and one in twenty for 1995. The task force based its estimates of the full effects of the embargo on the synthetics import market rather than on the entire energy market.[9] These limitations in the study resulted in a bias upwards of the estimated gain in social surplus resulting from the synthetic-fuel subsidy. The expected social benefit of embargo protection from the program ranges from $430 million to $1.2 billion, depending on whether the 350,000- or the 1-million-barrels-a-day synthetic program is carried out. This benefit is roughly counterbalanced by the social cost to the environment and society at large of $440 million and $1.1 billion, respectively (see Table 3). The conclusion is that, given the loss of economic welfare and the environmental degradation, the economy would lose more through the implementation of this program than it would gain in insurance against embargoes. In short, synthetic-fuel commer-

---

[8] Federal Energy Administration, *Project Independence Report* (Washington, D.C.: U.S. Government Printing Office, November 1974), Appendix, p. 291.

[9] Synfuels Interagency Task Force, *Recommendations*, vol. 2, pp. 117-19.

cialization programs are not a very effective way to protect against an embargo.

A relatively attractive alternative to a federal synthetic subsidy program is offered by government maintenance of an oil stockpile. Estimates of the cost of stockpiles vary, given different assumptions about price and quantity. Yet this alternative is much more amenable to estimation in terms of real resource cost than the open-ended financing programs. The Federal Energy Administration's *Project Independence Report* shows that a stockpile against a one-year, 1-million-barrels-a-day import interruption in 1985, if the world oil price is $11, will cost $6.3 billion over the ten years 1975–85.[10]

In addition to its certainty in terms of cost, the stockpile program would contribute to national security in the short-range period 1975–85, when this nation's import exposure is projected to be greatest. Synthetics and shale at the most optimistic estimate will contribute to security only toward the end of the period. Finally, implementation of a petroleum stockpile allows a delay in the production of shale and synthetics, thus reducing uncertainty as to long-run price and costs. Costs would be reduced even further if technological breakthroughs are made during the period. Synthetics and shale could still be produced if economically feasible in the early 1980s.

Since the cost-benefit study was completed, the government has initiated an oil stockpile program as a provision of the omnibus energy bill enacted in 1975. The stockpile program would seem largely to obviate a national security role for synthetic fuels before the year 1995. The President's task force acknowledged this point: "As a protection against loss of imports for up to several years, a storage program would apparently be less costly than attempting to avoid imports by rapid development of synthetic fuels production."[11] The task force noted that eventually the synthetics approach would be cheaper than stockpiling, but that the nation would be poorer if the government allocated resources to synthetics now in anticipation of 1995 benefits.

Some proponents of the subsidy proposals maintain that the programs will yield substantial nonquantifiable benefits, prominent among them:

- the international leverage or improved bargaining position that would accrue to the United States as a result of a synthetic-fuels program;

---

[10] FEA, *Project Independence Report*, p. 10.
[11] Synfuels Interagency Task Force, *Recommendations*, vol. 2, p. F5.

- the value in world and domestic leadership of an activist position;
- the possible impact on reducing the power of the OPEC cartel;
- the value to the United States of lower oil payments made by other importing nations.

The subjective nature of these benefits makes their evaluation difficult. Given the weight of other evidence cited in this chapter, however, these hypothetical gains are unlikely to outweigh the excess of measurable costs over measurable benefits.

Let us consider two of these nonquantifiable benefits. What leverage will the commercialization program give the consuming nations in negotiating international oil prices, and what impact would the program have on the strength of the oil exporters' cartel? At present the answer to these questions provides little justification for subsidizing a commercial synthetic-fuel program. The price of Iranian heavy crude oil was recently reduced from $11.49 a barrel to $11.40 a barrel in response to weak world demand. The Iranians hoped to stimulate consumption of their heavy oil, since expenditures for imports are exceeding revenues from oil.[12] The cost of Iranian crude, including $.60 a barrel for transport costs, is almost $2.50 a barrel cheaper than the current projected cost of domestic shale oil at $14.50 a barrel, assuming for purposes of comparison that it will cost $.50 a barrel to transport shale oil to the Northeast.

Although this higher-priced alternative may tend to cap any further foreign price rises, shale oil would seem to be a weak tool with which to negotiate lower prices. The actions of the individual producers show that the current cartel price may be too high, even if there were no alternative in the form of U.S. domestically produced synthetics. The current world price of oil may be so high that producers would increase their revenues by lowering prices, regardless of the availability of subsidized alternative U.S. domestic sources. There is some advantage, however, to knowing the maximum crude-oil price which the cartel can set in the future. If the world demand for petroleum were to increase, then the cartel's profit-maximizing price might become higher than the price of the U.S. domestic alternative, and the two advantages listed—bargaining power and impact on the OPEC cartel—would accrue. This factor, however, is already incorporated in the task force report. Much of the expected benefit

---

[12] "Iran Cuts Price on Crude Oil," *Washington Post*, February 16, 1976, p. A1.

of synthetic-fuel production results from imported oil prices being set at a lower level than without the program.[13]

Two further nonquantifiable benefits are claimed for the synthetic-fuels commercialization program. It is contended that there is leadership value in undertaking the program and goodwill value when other consuming nations make lower oil payments. The first benefit would probably accrue to the program if the United States were already undertaking every other economic measure to end the cartel arrangement of the Organization of Petroleum Exporting Countries. As long as the U.S. price is below the world price, however, this nation's policies appear to be strengthening the OPEC cartel. Any leadership value of a synthetic-fuels program is likely to be outweighed by the example set by our conspicuous consumption of energy.

The United States currently ranks last among the market-oriented industrialized countries in energy conservation following the January 1974 oil price increases.[14] Government energy-price-control policies are largely responsible for this by encouraging consumers to believe that oil is much cheaper than the cost of meeting increased demand—through oil imports. If domestic oil and natural gas prices were to be decontrolled, consumers would tend to demand a lower quantity at the higher prices and domestic producers would have further incentive to increase supplies. Not only would total oil demand fall, but also imported oil demand would fall faster.

The Impact of Decontrol of Domestic Energy Prices

The deregulation of domestic oil and gas prices, providing a rational alternative to a federal subsidy program, would involve substantial readjustments in the U.S. economy. Petroleum is used in a number of industries not directly related to energy, such as plastics and agricultural products. Rising petroleum prices would directly increase costs for these industries. The impact of initial cost increases might generate inflationary expectations and touch off a new round of inflation. Certainly the decontrol of petroleum prices would lead to some "windfall profits."

The possible inflationary impact of the program must be seriously considered. It is important to distinguish, however, between the impact of foreign price hikes and a rise in prices which would be

---

[13] Conversation with Steve Tani, Stanford Research Institute, consultant on the synthetic-fuels task force report.

[14] Leo Ryan, "U.S. Gets Lowest IEA Rating," *Journal of Commerce*, December 1, 1975, p. 28.

associated with domestic decontrol. Foreign price increases, in the face of declining domestic output, lead to a redistribution of income from American consumers to income earners in foreign nations. Domestic decontrol would lead to a redistribution within the United States. Furthermore, the initial "windfall profits" of domestic producers would largely be dissipated both through payments to manufacturers of drilling equipment and petroleum engineers and in the form of higher tax payments.

Decontrol will lead to an expansion of domestic oil production which will in turn stimulate other related domestic industries. Moreover, the impact of decontrol on prices will be minimized if the program is carried out in stages which allow and encourage the public to plan and implement methods of economizing on energy, before the full impact of decontrol is felt.

It is sometimes argued that decontrol would strike hardest at the poor. The evidence does not show lower-income classes spending larger proportions of their incomes on energy than higher-income classes. If high energy prices to the poor become a concern, however, direct assistance could be employed. Public assistance payments could be increased, and the recipients themselves could decide how much they wanted to spend on relatively higher-priced energy and how much they wanted to spend on other relatively lower-priced items.[15] All income groups would continue to have an added incentive to conserve energy.

Finally, the impact of decontrol would strengthen the U.S. balance of payments as consumers shifted to domestic fuels. The resulting expansion in the domestic money supply, coupled with the linkage effects between an expanding domestic petroleum and natural-gas industry and other sectors, could actually stimulate the economy in the long run. The impact of decontrol would be greatly influenced by such factors as the level of unemployed resources at the time of decontrol, the impact of expectations generated by allowing oil prices to rise, and the reaction of foreign suppliers.

Other Considerations

The Synfuels Task Force report is a serious attempt to assess the probable impact of a synthetics program. It relies, however, on assumptions which may tend to inject a favorable bias to the analysis. For example, the demand for energy in the United States is set forth as a given or predetermined factor in the task force model. The report

---

[15] Mitchell, *U.S. Energy Policy*, p. 2.

assumes that the demand for energy will continue to grow at 2.9 percent a year, close to the long-run growth rate of the gross national product. On the other hand, since the price of energy has risen significantly relative to that of other products, the demand for energy should begin to grow at a lower rate as consumers learn how to economize on a relatively more expensive factor. Thus, the demand for synthetic fuels in 1985 may be smaller than the task force assumes.

The task force forecast used a time horizon of forty years, a period long enough for significant adjustments to take place. The assumption of separate markets for liquid and gaseous fuels may prove untenable. Technological changes and relative price movements could completely eliminate gaseous fuels from the markets, if the price of their marginal source rises above the price of the marginal source of liquid fuels. Thus, the task force's assumption may suggest an artificially high demand for synthetic gas which increases the projected benefits of a synthetic commercialization program.

Supply considerations also present a significant degree of uncertainty. Some experts estimate that the 1975 level of crude-oil production could be maintained for thirty-five years with a marginal extraction cost equal to or less than the price of comparable oil shale, while others estimate that the 1975 level could be maintained for sixty-five years before marginal extraction costs exceeded the cost of shale oil.[16] Doubts also exist about how long the government will continue to regulate domestic prices of crude oil and natural gas. The resolution of this uncertainty does not rest with a government subsidization program, but rather with a removal of federal interference with the domestic market.

Conclusions. This chapter has examined the impact of synthetic-fuel subsidy programs on consumers. The effect of the programs is to force consumers, through governmental action, to utilize high-cost energy sources. The President's task force found that the costs to the nation outweigh the benefits to national security of the program. In fact, the insurance provided by the program against oil embargoes would be balanced by the environmental costs. Costs will outweigh benefits even more heavily if we take into account the probable lower growth in demand rather than the task force estimate. Another likely development which would decrease the cost-benefit ratio is a higher conventional fuel supply resulting from oil and gas price decontrol, or the buildup of an energy stockpile.

---

[16] The figures in this section are estimates based on Synfuels Interagency Task Force, *Recommendations*, vol. 2, p. 23, Figures 12 and 13.

4

ANALYSIS OF PROPOSED ENERGY INDEPENDENCE AUTHORITY

This chapter is devoted to an examination of the specific provisions of the Energy Independence Authority Act (S.2532). The bill received strong support from the Ford administration, but the 94th Congress failed to enact it. Nevertheless, a careful examination of the bill seems warranted by the likelihood that its various provisions will reappear in future legislation. The impetus toward such measures has become a virtual constant in U.S. political life.

Financial Assistance by EIA

The proposed statute provided a broad range of powers and discretion to a new federal agency, the Energy Independence Authority (EIA). The EIA would have been empowered to provide "financial assistance to any business firm" deemed to meet the requirements of the act.

Numerous forms of financial assistance were authorized under the act, including advances, extensions of credit, investments, participations, loan guarantees, price guarantees, purchases and lease-backs of facilities, and purchases of convertible or equity securities. Subject to specific limitations in the act (described below), the EIA could have provided the financial assistance "in its sole discretion and upon such terms and conditions as it may determine." It would have been specifically authorized to make "high-risk" loans.

The financial assistance could have been used in a variety of ways: to buy or build productive facilities, to acquire factories, equipment, and supplies, to take over and develop land and mineral rights, to purchase services, and to provide working capital. In his transmittal message to the Congress, President Ford referred to creating a "new partnership" between the private sector and the federal government on "vital" energy projects.

31

Projects would have been given financial assistance if the EIA board of directors determined they would make a "significant" contribution either to the achievement of energy independence by the United States or to the long-term security of energy supplies for the United States. The projects would have to be shown unable to receive "sufficient" financing upon commercially "reasonable" terms from other sources to become commercially "feasible." No definitions were provided to guide the EIA in determining the meaning of the terms *significant, sufficient, reasonable,* or *feasible.*

A set of general technical criteria was provided to guide the EIA board of directors in choosing projects:

- projects that use or stimulate the application of technology essential to developing, producing, transmitting, or conserving energy and not in widespread domestic commercial use.
- projects that use or stimulate the application of technology essential to producing or using nuclear power.
- projects that use or stimulate the application of technology for generating or transmitting electricity from fuel sources other than oil or natural gas.
- projects that utilize existing technology in widespread commercial use but either (a) are so large that they would not be undertaken without help from EIA, or (b) involve an institutional or regulatory arrangement not in widespread domestic commercial use that could lead to improving the development or production of energy or innovative transportation or transmission facilities.
- projects that use or stimulate the application of technology for protecting the environment in any of the previous types of projects.

According to the fact sheet issued by the White House on the EIA proposal, the projects that could have been supported ranged across the full spectrum of energy, excluding research. They included commercialization of such synthetic-fuel technology as coal gasification, liquefaction, and production of oil from shale, as well as solar and geothermal energy technology. In addition, such conventional technologies as uranium enrichment, coal, nuclear, and geothermal power plants would qualify. Projects of unusual size or scope indicated in the bill included new energy parks and major new pipelines for transporting oil and gas.[1]

---

[1] Office of the White House Press Secretary, *Fact Sheet, Energy Independence Authority,* October 10, 1975.

The Energy Independence Authority would have been authorized to sell $25 billion of capital stock to the secretary of the Treasury, subject to the availability of congressional appropriations. EIA could also issue its own notes, debentures, or bonds, up to a total of $75 billion outstanding at any point. The Treasury department would be authorized but not directed to buy these securities. All EIA contractual commitments to provide financial assistance would be general obligations of the United States backed by its full faith and credit. As a result, the new agency might have as much as $100 billion available to it at any given time. This amount can be compared with the total of $312 billion in outstanding loans by all federal departments and agencies as of June 30, 1975.

Regulatory Participation by FEA

Because regulatory problems often make financing difficult by adding uncertainty about an energy project's ultimate approval, the proposed EIA statute gave the Federal Energy Administration substantial power to intervene in the proceedings of government regulatory agencies. FEA would have acquired the authority to issue a composite license application which would be the sole application required by all federal agencies in proceedings related to an energy project. The standardization was designed to eliminate the delays a utility encounters in securing permission from a variety of government regulatory agencies. One electric company recently was required to obtain twenty-four different approvals from thirteen government agencies prior to building a new generating plant.[2] FEA would also have been empowered to request a federal regulatory agency to reconsider a decision, or itself to join in an appeal on the part of the applicant. The bill required such a petition to be acted on within thirty days, but FEA could not override or bypass existing regulatory agencies or their proceedings.

FEA could certify that an energy project—even one not receiving aid from EIA—was of "critical importance" to achieving the purposes of the EIA act. Federal regulatory agencies would have been authorized (but presumably not required) to give preference in their review processes to projects receiving such a designation. The act called for "diligent efforts" by the regulatory agencies to render a decision in eighteen months, or a shorter period if FEA specified "for good cause." This preferred treatment was intended only to speed the decision-making process and not to deal with the substantive issues on which the regulatory agency was ruling.

[2] Weidenbaum, *Government-Mandated Price Increases*, pp. 92-93.

The bill also provided that judicial review of a federal regulatory agency's final action on an energy project certified by FEA took precedence on the docket over all cases except those considered of "greater importance" by the court. The critical projects must be scheduled for hearing and trial at the earliest "practicable" date and expedited "in every way."

The present study is not the appropriate place for a general review of the effectiveness of environmental and other federal regulatory programs. The point to be made is that the recent expansion of regulatory legislation into ecological and other social areas has resulted in a new kind of delay in carrying out large new developmental projects in the United States. With few exceptions, the regulatory process delays each proposed new energy development project for an unpredictable period and at some significant cost. The EIA type of credit proposal does not face the issue, but merely initiates a new federal effort to overcome, at least in part, the adverse effects of an earlier federal effort. The Synfuels Interagency Task Force, after listing the various regulatory requirements, stated the issue succinctly: "In summary, some of these requirements could easily hold up or permanently postpone any attempt to build and operate a synthetic fuels plant." [3]

The more significant regulatory constraints include the following:

- Preparing an environmental impact statement, as required by the National Environmental Policy Act of 1969.
- Meeting new source performance standards for air quality, under the Clean Air Act Amendments of 1970.
- Meeting the hazardous pollutant emission standards, under the Clean Air Act Amendments of 1970.
- Meeting the state air-quality implementation plans required by the Clean Air Amendments of 1970.
- Obtaining necessary point source discharge permits, under the Water Pollution Control Act Amendments of 1972.
- Meeting state water-quality standards and water-quality management plans, as promulgated under the Water Pollution Control Act Amendments of 1972.
- Complying with limitations applicable to "underground injections," under the Safe Drinking Water Act of 1974.
- Complying with the regulation of interstate pipeline transmissions, under the Interstate Commerce Act.
- Complying with the prohibition against a carrier transporting its own products, under the Interstate Commerce Act.

[3] Synfuels Interagency Task Force, *Recommendations*, vol. 1, p. 134.

- Complying with the allocation of railroad cars transporting coal, under the Interstate Commerce Act.
- Complying with the regulation of interstate transmission of synthetic gas once it is mixed with natural gas, under the Natural Gas Act.
- Obtaining necessary plant and mine leases, from the U.S. Bureau of Land Management.
- Obtaining necessary water allocations, from the U.S. Bureau of Reclamation.
- Complying with the Coal Mine Health and Safety Act of 1969.

The comments of the Synfuels Interagency Task Force on the impact of the various regulatory requirements deserve far more attention than they have received to date. The task force evaluated the effects of the environmental impact statements (EIS) required by the National Environmental Policy Act of 1969 (NEPA) as follows:

Thus, the major uncertainty under NEPA is not whether or not the project will be allowed to proceed, but rather the length of time it will be delayed pending the issuance of an EIS that will stand up in court. The cost of such delays (construction financing and inflated raw materials and labor costs) is an obvious potential hazard to any synfuels project. . . .

In summary, the cost and delay occasioned by NEPA constitute a substantial disincentive, aggravated by the fact that in dealing with new processes it is very hard to anticipate what the EIS requirements will be and on what grounds the EIS may be attacked. The general guidelines offered by the Council on Environmental Quality (40 CFR Part 1500) provide a drafting framework but no assurance of compliance.[4]

Restrictions on EIA

The proposed law established numerous criteria for EIA decision making. Some of these were clear restrictions, but others represented the general advice or vague wishes of the sponsors and could be met by merely reciting stock phrases. In the aggregate, these restrictions were likely to slow down considerably EIA's speed of action as well as to require substantial paperwork to justify the measures it did take.

---

[4] Ibid., pp. C-18 and C-19.

The following restrictions on financial assistance were set forth in the proposed bill.

The projects assisted were limited to those that would not receive "sufficient" financing upon commercially "reasonable" terms from other sources to make the project commercially "feasible." According to Federal Energy Administrator Frank Zarb, this provision was designed to provide maximum encouragement for private lenders to participate in energy projects.[5] Upon reflection, this provision was not likely to operate independently of other federal energy actions (or inactions). For example, so long as federal price controls remained on oil or natural gas, or other price uncertainties continued, a very substantial portion of proposed new energy projects would be considered uneconomical by private lending sources, and thus would meet the standard for EIA aid. Yet elimination of price controls and a return to a free-market situation would result in a far greater availability of private credit for the very same projects and thus in a sharply reduced need for EIA financial assistance.

Financial assistance was to be provided in a manner which, "to the extent possible," would not enhance "unduly" the recipient's competitive position. This objective was to be achieved by the requirement that EIA aid be given on terms comparable to those available in private credit markets. The purpose appeared to be equitable treatment, but in actuality the very nature of EIA projects worked against equity. Projects submitted to EIA would not have received adequate private funds precisely because they were considered too risky at normal market rates of interest. The implicit subsidy could be substantial; it would be the difference between the rate of interest paid to EIA (the normal market rate of interest) and the higher rate required to obtain private finances for the project.

Financial assistance could not be in the form of grants-in-aid. That is, EIA could not make outright donations or gifts.

"Adequate provision" must be made to insure that EIA would share in the profits that might result "commensurate" with the risk that it assumed. No guidelines were set forth as to what would be "adequate" recompense for the risk assumed by EIA, or how that risk would be measured. An analysis submitted to the Congress by a senior partner of the investment banking firm of Dillon, Read and Company was relatively pessimistic as to the availability of private

---

[5] Office of the White House Press Secretary, Press Conference of Frank Zarb, Administrator of the Federal Energy Administration, James T. Lynn, Director of the Office of Management and Budget, L. William Seidman, Assistant to the President for Economic Affairs, and Robert Fry, Deputy Administrator of ERDA, October 10, 1975, p. 4.

capital for synthetic-fuel projects in the absence of government assistance: "In order for synthetic fuel projects to successfully compete for capital funds, we believe that prior to the commencement of construction *assurance must be given to potential lenders* that all funds necessary to complete the projects have been committed and *their loans will be repaid under all circumstances. . . ."* [emphasis supplied].[6] It appears that 100 percent guarantees by the federal government were envisioned, in at least some cases. The risk involved for the private capital committed under such circumstances would be nominal.

Financial assistance could be given to regulated companies (for example, electric utilities) only if the state or local regulatory agency had met several stated conditions. In the aggregate, meeting those conditions would involve a basic loss of authority on the part of state or local regulatory agencies.

- The state or local regulatory agency would have to issue a certificate of necessity for the project as prescribed by EIA.
- It would be compelled to enter into a three-party agreement with EIA and the regulated company requiring the regulating agency to permit, without prior hearing, quarterly rate adjustments to yield sufficient earnings for a "minimum level" of coverage of interest charges. This measure would constitute a major change in state regulatory practices, and the assumption by EIA of a major responsibility—determining the allowable rate of return to be earned by a regulated utility.
- EIA would establish what that minimum interest coverage should be. It would be required to set interest at a level sufficient to assure repayment of its investment and restoration of the regulated company's credit rating to a level at which it could raise its own capital at "favorable" interest rates.

Financial assistance could not be given to projects involving technology in the research and development phase. Supposedly those activities come under the jurisdiction of the Energy Research and Development Administration.

Financial assistance could not be given to projects if the applicant did not display "satisfactory" levels of "efficiency, management capacity, or similar factors." These factors, always difficult to evaluate, are customarily considered by private sources of financing before making an investment decision.

---

[6] Statement of Arthur B. Treman, Jr., p. 5.

No project could be approved unless EIA had taken into account "competitive alternatives" to meet the same energy need. No guidance was given as to how EIA was to take those alternatives into account, other than noting their existence. Coupled with the other restrictions, it would seem that the EIA review process could have become expensive and time-consuming.

Financial assistance to any one business concern or affiliated business concerns was limited to $10 billion. This provision was designed to assure that no solitary enterprise or single group of companies would receive the bulk of the aid and that smaller businesses would participate in the program. Yet the $10 billion maximum does appear generous.

Before making any commitment to extend financial assistance, EIA must seek the advice of the members of the Energy Resources Council and of any other federal agency designated by the President. The advice was to assist in determining whether the financial assistance would "further the purposes" of the EIA law and how the assistance related to other programs and national policies. Such advice was required to be provided to EIA within thirty days.

"To the extent practicable" and "in the judgment of the Board of Directors," financial assistance was to be in the form of loans and loan guarantees, rather than equity investment or price guarantees.

The proposed statute also provided numerous general and specific guidelines to EIA. For example, the interest rate that EIA charged on its loans would have to be at least equal to that paid by creditworthy borrowers to private lenders on "comparable" terms (other than interest rate) for projects of a similar nature. No upper level was set on the interest rates that EIA could charge. EIA's board of directors was required to give consideration to the "needs and capacities" of the recipient, the prevailing rates of interest (public and private), and the agency's need to sustain continuing operations out of returns on investment. EIA would also be required to charge "reasonable" fees for issuing guarantees and for making commitments to provide other forms of financial assistance.

At various points, the proposed law referred to the need to attain energy independence in a manner "consistent" with the protection of the environment and giving "due regard" to the need to protect the environment. All business concerns receiving financial aid from EIA would also be required to meet federal labor standards and equal employment opportunity requirements, presumably in all of their activities and not just those receiving EIA assistance. All laborers and mechanics employed by contractors and subcontractors in any construction, alteration, or repair work on EIA-funded projects must receive at least the "prevailing" wages, under the Davis-Bacon Act.[7]

The proposed act stated a general concern for a competitive private-sector economy. It spoke of providing financial assistance in a manner that preserved "economically sound and competitive" industry sectors, while minimizing any economic "distortion or disruption of competitive forces," though it did not indicate how this was to be done. Another section of the act stated as its purpose to supplement and encourage—rather than compete with—private capital investment and activities, "recognizing that the private sector must play the primary role." In comparison with the specific powers given to the new government agency to make business investments, these statements seemed mere "window dressing." Another section stated that the "maximum" amount of financing from sources other than EIA, preferably private sources, should be sought in connection with any project for which financial assistance was provided. As in most of the other provisions of this nature, the EIA board of directors was to judge whether this vague requirement was met.

Budgetary Impact

The establishment of EIA would have represented a major expansion of the category of "off-budget agencies." In the fiscal year 1975, approximately $10 billion of outlays (net of receipts) were made by the existing agencies in this category, a subterfuge (as explained in Chapter 2) for minimizing budget totals. The revenues and outlays of the Energy Independence Authority would have been excluded from the totals of the federal budget and exempt from any spending ceiling set by Congress. Likewise, the Treasury department's purchases and sales of the capital stock of EIA, as well as the dividends that it received, would not be included in the budget. The one exception is that EIA net earnings or losses would be included in the budget totals. Otherwise, the Energy Independence Authority would be treated as an off-budget agency, with its revenues and outlays not reflected in the federal budget. But the total of Treasury department, and thus aggregate federal, borrowing from the public would of course include the transactions of EIA. Employees of EIA, including full-time directors, would be considered employees of the U.S. government for purposes of eligibility for benefits related to employment.

---

[7] John P. Gould, *Davis-Bacon Act* (Washington, D.C.: American Enterprise Institute, 1971); Armand J. Thieblot, Jr., *The Davis-Bacon Act* (Philadelphia: University of Pennsylvania, The Wharton School, 1975).

The President's budget for the fiscal year 1977 contained an estimate of $650 million of net outlays by EIA ("off budget") and $42 million of net losses, included in the total of budget expenditures. The budget message also stated that, pending enactment of the EIA, the Ford administration strongly supported the immediate authorization of a synthetic-fuels commercialization program to be administered by the Energy Research and Development Administration. The new budget included $503 million in budget authority for the fiscal year 1976 to cover $2 billion in loan guarantees as a first step in implementing the new program.[8] Why a guarantee program required this magnitude of budget authority was not made clear, unless substantial losses were expected to be charged to the budget. Approximately $4 million of such net outlays were budgeted in the fifteen-month period, July 1, 1975, through September 30, 1976.

The budget message also assumed that, with the creation of EIA, the synthetic-fuels program would be transferred to the new agency during the fiscal year 1977 and then grow to a level of $6 billion in loan guarantees. The synthetic-fuels promotion program may well have been the entering wedge for broader legislation along the lines of the basic EIA proposal. In late 1975, Congress seriously considered a similar proposal, but the House of Representatives finally rejected it in a floor vote. The House Science and Technology Committee held hearings on a comparable bill in the spring of 1976.[9]

The practical problems of incorporating the initial synthetic-fuel credit program into the EIA could have been formidable, in view of the numerous procedures that would have to be followed before making a loan guarantee.[10] Following are the restrictions that would have been placed on the synthetic-fuel program, many of which differ in detail from the procedures specified in the EIA bill:

- The Bureau of Competition of the Federal Trade Commission was to review each proposed loan guarantee. The FTC is to give "serious and meaningful attention" and provide a "comprehensive and adequate" response.
- The Department of Justice was to make a similar review.
- A report was to be submitted on each proposed guarantee to the House Science and Technology Committee and the

---

[8] *Budget of the United States Government for the Fiscal Year 1977* (Washington, D.C.: U.S. Government Printing Office, 1976), p. 87.

[9] U.S. Congress, House, *Conference Report to Accompany H.R. 3474*, Report No. 94-696, 94th Cong., 1st sess., 1975; House Committee on Science and Technology, *Synthetic Fuel Loan Guarantees*, vol. 1.

[10] Ibid., pp. 49-58.

Senate Interior and Insular Affairs Committee. Either chamber could then disapprove proposed guarantees (of $350 million or more) within ninety days.

- Concurrence of the Department of the Treasury was required on the timing, interest rate, and "substantial terms and conditions" of each guarantee.
- The administering agency was to be sensitive to the congressional concern that concentration in the energy business would not be "further aggravated" through the loan guarantees.
- Guarantees were limited to construction and start-up costs.
- A "high priority" was assigned to the demonstration of the synthetic production of pipeline quality gas.
- If the administering agency sought to override the negative recommendation of a governor, the burden was on the agency to show that the particular facility was indeed in the national interest.
- No oil shale commercial demonstration facility receiving a loan guarantee could be larger than necessary to demonstrate the commercial viability of the process.
- The agency was to have due regard for the need for competition in making guarantees.
- It could require each new commercial project to cover the capital costs for essential public community facilities. (Where the private project could not adequately provide for the capital costs of new community facilities, ERDA could make direct loans and could forgive all or part of their repayment.)

Some of the proponents of the limited synthetic-fuel loan guarantees were staunch opponents of the larger government credit subsidies for energy. In his testimony favoring the $6 billion program cited in Chapter 2, Professor Henry D. Jacoby made this point quite forcefully:

If I thought this bill was a prelude to a massive program of loan guarantees for new energy facilities, for multiple plants with known technology and not just for a limited set of demonstrations, then I would oppose it. I think it would be a terrible mistake to embark on a large-scale program of hidden subsidies for energy supply from new, capital-intensive technologies. . . . The disadvantage of the widespread use of loan guarantees is that they will obscure the true cost to the economy of new energy sources or energy-saving technologies. They will hide the cost from policy-makers, and thereby bias decisions in this realm. More important, they hide the true cost from consumers and encourage wasteful consumption practices.[11]

Administration of the Agency

The basic power of the Energy Independence Authority was to be lodged in a five-member board of directors appointed by the President with the advice and consent of the Senate. The members would serve at the pleasure of the President who would name the chairman and set the compensation of all of the board members. The position of chairman was to be a full-time one, while the others might be part-time. Not more than three of the members could be of the same political party, and no qualifications were established for board members other than U.S. citizenship.

The chairman of the board, designated the chief executive officer of EIA, would be responsible for its management and direction, including administrative expenditures. The chairman, in turn, could appoint and set the salary of all of the employees of EIA. One hundred "supergrade" (GS-16, 17, and 18) employees would be authorized, in addition to any that might be assigned to EIA out of the pool established by existing law. Twenty-five of the new supergrades could be "political appointees," that is, they could be hired without regard to the civil service statutes governing "classification and appointment." The chairman could specifically appoint a "reasonable" number of additional "executive officers" under employment agreements not exceeding five years.

Although EIA was to avoid acquiring permanent controlling interest in commercial activities, the law contemplated that EIA might wind up in the energy business by operating some of the private projects it financed. The agency could take over any collateral that it accepted as security for its loans. When EIA did acquire control of operating assets prior to the commencement of their commercial use, it was limited in retaining them to two years. If control was acquired by foreclosure or pursuant to a default under a lease, EIA could retain control of operating assets for as long as four years.

The agency would have had a broad grant of legal immunity. No private individual or organization or state or local government could sue EIA for taking actions inconsistent with its statutory charter, or for neglecting to discharge its duties under the act. Only the

---

[11] Statement of Dr. Henry D. Jacoby, *Synthetic Fuel Loan Guarantees*, vol. 1, pp. 370-71.

attorney general could institute such a suit. Private parties would be free to sue EIA for breach of contract, and the agency's activities would come under the Federal Tort Claims Act. EIA was also to be given broad access to information about the operations of applicants for financial assistance. Each applicant was to consent to whatever financial examination EIA required and to provide any reports of examinations by "constituted authorities."

EIA's issuance of its securities, as well as loan guarantees or other obligations which have a direct impact on capital markets, was to be subject to approval by the secretary of the Treasury as to timing, method, interest rate, and other terms and conditions. At the discretion of the secretary, EIA obligations would be channeled through the Federal Financing Bank. EIA was authorized, but not required, to deposit funds with the U.S. Treasury and, with the approval of the secretary of the Treasury, in any Federal Reserve bank.

The secretary of the Treasury was to make an annual determination of the minimum dividend rate to be paid by EIA on its outstanding capital stock. He was to take into account the current yield on marketable U.S. obligations, but otherwise had wide discretion in the matter. EIA's board of directors, in turn, could set the dividend rate at the current level or higher, at their discretion. The board could defer paying dividends if it determined that the funds should be used instead to provide financial assistance, or if no funds were available. In such event, EIA would pay interest on the unpaid dividends at a rate determined by the secretary of the Treasury. In making that determination, the secretary again was to take into consideration the average yield on marketable obligations of the United States.

The interest rate on the bonds that EIA sold to the Treasury was to be at least equal to a minimum rate set by the secretary, taking into consideration the average yield on outstanding marketable obligations of the United States of comparable maturity. The interest payments could be waived at the secretary's discretion, but EIA was to pay the Treasury interest at the designated rate on the unpaid interest that was due.

The proposed statute establishing the EIA contained several other interesting provisions. Section 204 waived the government's normal immunity from taxation; it made its real property subject to federal, state, and local taxation. Thus, rather than the modest "payments-in-lieu" of taxes that some federal agencies make, often on a voluntary basis, EIA property was to be subject to the same taxes as privately owned property. The waiver of federal tax immunity was even broader in the case of commercial activities that EIA came to own.

Any company or activity acquired or established by the agency that produced, distributed, or sold energy, fuel, or related products would be fully liable for all federal, state, and local taxes, just as if it were privately owned. EIA could sell in public or private transaction the stock, bonds, capital notes, or other financial instruments that it acquired through default, or otherwise.

The President of the United States would have been authorized, but not required, to appoint an advisory panel to study the effects of EIA's issuance of obligations and provision of financial assistance on the functioning of the nation's capital markets. The panel could include in its study the effects of EIA activities on the volume and distribution of capital flows to and within the "energy development sector" of the economy. The President would set the duration, organization, membership, and scope of the panel.

EIA was to retain one or more firms of "nationally recognized" public accountants to prepare an annual audit of its accounts. The U.S. General Accounting Office would also have been authorized to audit EIA accounts. In a relatively new control over lobbying activities written into the bill, EIA was to record the written and other communications from private individuals and public officials expressing an opinion on any proposal for financial assistance. The agency was to make such records available to the public upon request.

The proposed law contemplated that EIA would stop making new commitments for financial assistance by June 30, 1983, and would give no new financial assistance after June 30, 1986, when it was scheduled to be terminated.

Anticipated Results

The official statements issued by the Ford administration in support of the proposed Energy Independence Authority were vague on the specific results to be expected from the new agency's activities. According to the White House fact sheet on EIA, the Federal Energy Administration estimated that investments for energy independence could total about $600 billion (in 1975 dollars) over ten years.[12] In a press conference in October 1975, Federal Energy Administrator Frank Zarb was more specific. He stated flatly, "It is our estimate that it is going to require $600 billion over the next ten years to become independent. . . ."[13]

---

[12] Office of the White House Press Secretary, *Fact Sheet*, p. 2.
[13] Press conference of Frank Zarb, p. 2.

In forwarding the EIA proposal to Congress, President Ford maintained that the risks in many of the suggested projects were so great that private capital markets would not provide necessary financing. "The uncertainties associated with new technologies inhibit the flow of capital," he stated.[14] Thus, assuming that its full authority were used, the EIA might have assisted—through loans, guarantees, equity purchases, or in other ways—as much as one out of every six dollars invested in domestic energy over a decade. The White House contended that many new projects, such as uranium enrichment plants, were too large and economically risky to be financed by the private sector alone. Also, emerging technologies, such as solar energy and shale oil, involve long lead times and technological uncertainties, as well as considerable risk if world oil prices dropped.

In terms of direct contribution to energy independence, the White House fact sheet offered the hopeful statement that the $100 billion of EIA financial aid "could help assure" that the equivalent of "up to" 10–15 million barrels of oil a day of new energy production would be realized in 1985. No direct connection was made by the fact sheet between any specific amount of financing and the resultant increase in domestic energy production. In its annual report, EIA would evaluate the contribution, past and future, of each project or activity that it assisted in fulfilling the purposes of the EIA act, including "where possible" a precise statement of the amount of domestic energy produced or to be produced.

It is interesting to note that the privately financed Alaska pipeline project is expected to yield 1.2 million barrels of oil a day by November 1977. The Alyeska Pipeline Service Company estimates the total investment cost at something over $7 billion—a substantial amount, but a relatively modest price tag compared with EIA's $100 billion.

The fact sheet gave no indication why the round figure of $100 billion, rather than $50 billion or $150 billion, was needed to achieve the independence objective by 1985. Nor was any reason offered for 1985 as the target date. Moreover, no attention was given to the contribution of other programs and methods dealing with the energy problems facing the United States. For example, greater reliance on the price system would encourage domestic production and at the same time dampen domestic demand for energy. Creating an emergency petroleum reserve would help to insulate the domestic economy

---

[14] Office of the White House Press Secretary, *Text of Letters from the President to the Speaker of the House of Representatives and the President of the Senate,* October 10, 1975.

from the threat of another OPEC oil embargo and would reduce the need to achieve absolute energy independence.

In the opinion of a number of analysts, the most attractive way of increasing domestic energy production is to concentrate on the unexplored provinces of the outer continental shelf and Alaska. According to Professor W. N. Peach of the University of Oklahoma, the most promising area for development for the United States, at least in the short run, lies in the vast deposits of oil and gas in the outer continental shelf. He cites estimates of the U.S. Geological Survey that the potential resource base in the OCS is 1–1.5 trillion barrels of oil and 3–4 quadrillion cubic feet of natural gas, from which the petroleum industry should be able with current technology to recover 160–190 billion barrels of oil (about twice as much as the industry has produced in its entire history) and from 800 trillion to 1.1 quadrillion cubic feet of gas.[15] Arlon R. Tussing, professor of economics at the University of Alaska and consultant to the Senate Committee on the Interior, echoes this view: "It is preposterous even to consider a major effort to produce new kinds of fuels from investments that cost ten, twenty or thirty thousand dollars per daily barrel of oil equivalents, and take a decade or more to complete, while we have yet to drill 95 percent of the sedimentary acreage on the Outer Continental Shelf." [16]

Professor Tussing estimates the investment cost per daily barrel of oil from the outer continental shelf at nearer to $5,000 than $20,000 and the payoff time at a fraction of that required for the synthetics. He also considers it "preposterous" to consider manufacturing synthetic gas or moving it by pipeline 5,000 miles at a cost of $3,000–$5,000 per thousand cubic feet while price regulations still effectively prohibit producers from looking for or developing conventional natural gas that would cost little more than 52 cents per thousand cubic feet.[17]

Decontrol of oil and gas prices would help to create an environment conducive to increased exploration. Petroleum companies are reported to be reducing their development budgets. Seismic exploration, an indicator of future drilling activities, has been declining since the middle of 1975.[18] One industry executive, Theodore R. Eck of

---

[15] Peach, *The Energy Outlook for the 1980's*, p. 12.

[16] Arlon R. Tussing, *Good and Bad Examples in the Search for Energy Independence*, prepared for the Rocky Mountain Energy-Minerals Conference, Billings, Montana, October 16, 1975, p. 12.

[17] Ibid.

[18] Sanford Rose, "Why Big Oil Is Putting the Brakes On," *Fortune*, March 1976, p. 110.

the Standard Oil Company of Indiana, has stated: "Given the uncertainty over the future price of oil and gas, only an extreme gambler would be increasing his exploration activity at the present time. This company plans to drill 35 percent fewer exploratory wells in 1976 than in 1975."[19]

Revisions in government policy regarding the leasing of offshore lands also could encourage a faster rate of exploration of conventional energy sources. One method would be to reduce or eliminate "front-end" payments for such leases and rely more heavily on royalty payments on actual output. The resultant improvement in cash-flow prospects might turn some potentially negative prospects into more promising ones.

Another area of uncertainty is created by the fear of the major vertically integrated oil companies that the government will force them to split up into smaller units ("divestiture"). According to Professor Richard B. Mancke of Tufts University, ". . . the mere threat of divestiture discourages oil companies from making investments of the magnitude necessary if the United States is to reduce its oil import dependence."[20] Forced divestiture would only make it more difficult or at least more expensive for an individual company to amass the large pool of capital which is required for many new energy projects, conventional or synthetic.[21] Offsetting effects of a federal divestiture effort by means of a new federal financing subsidy (such as the EIA) seems analogous to pressing simultaneously on the brake and the accelerator of an automobile.

[19] Ibid., p. 11.
[20] Richard B. Mancke, "Competition in the Oil Industry," in Edward J. Mitchell, ed., Vertical Integration in the Oil Industry (Washington, D.C.: American Enterprise Institute, 1976), p. 71.
[21] See Edward J. Mitchell, "Capital Cost Savings of Vertical Integration," in Vertical Integration, pp. 73-103.

5

FINDINGS AND
CONCLUSIONS

It is difficult to forecast the course of development of a government program such as the proposed Energy Independence Authority, which was not enacted by the 94th Congress and may never come into being. Yet this initial survey indicates a variety of potential shortcomings of the proposals for federal credit subsidies to promote the development of domestic energy production. These problem areas range from a massive expansion of the role of government in the private sector, to a proliferation of off-budget financing, to the basic question of the effectiveness of subsidies in promoting energy development.

The Basic Question: Why a Federal Subsidy?

Why does the domestic energy industry need federal credit assistance in the first place? The size of the undertaking in itself does not necessitate government assistance; large commercial energy projects, such as the $7 billion Alaska pipeline project, are proceeding with private financing. Nor is it a question of the weak financial condition of the energy industry. Although public opinion exaggerates the major oil company profits, the industry's rate of return over the years is just about average for manufacturing companies,[1] while its asset structure is strong. Neither is it a question of a static or declining demand for energy. Individual estimates vary, and for good reason, yet every forecast of future energy consumption in the United States shows a rising trend,

---

[1] James M. Dawson, *Windfall Profits?* (Cleveland: National City Bank, 1974), pp. 2-3.

far beyond the capacity of existing domestic energy supply sources.[2] Why then does the private sector display such limited interest in the development of new domestic energy sources?

The answer is clear: under present circumstances, many such undertakings either are uneconomical or are restricted by federal regulatory programs. The cost of competitive, conventional sources of energy is generally much lower than that of potential alternatives. The situation is likely to change as the result both of basic economic forces (if we let them work) and of the adoption of more sensible regulatory policies.

In the years to come, as more marginal and thus higher-cost conventional energy supplies are used, and as these fuels become relatively scarce, the gap is likely to narrow between the cost of fuels from existing sources and the cost of new alternatives. Products such as synthetics could become competitive without government subsidy. The November 1975 report of the Synfuels Interagency Task Force stresses this point:

> The results of the analysis imply that under normal investment and risk circumstances, market forces are likely to cause the introduction of synthetic fuels in the 1985–1995 time period. With the right combination of prices and costs, production of synthetic fuels in 1995 might be as high as 9 million barrels per day although the expected average is 5 million barrels per day.[3]

Yet it must be emphasized once more that the price of conventional energy in the United States is being kept artificially low by government policy. Of course we all prefer to pay less for something rather than more, but short-run personal economy is hardly a guide to sensible public action.

Clearly, the lower the price of existing fuel sources is kept, the less attractive becomes the prospect of developing new domestic energy alternatives. An impartial observer can only gaze in wonder at an approach to policy which first keeps conventional fuel prices artificially low (via government controls) and then finds that new domestic energy sources will not be developed on a sufficiently large scale without special government assistance. The situation is made worse by the uncertainty arising from the limited duration of existing price controls. Given higher world prices, the controls give domestic

---

[2] FEA, *Project Independence Report*; Hausman, "Project Independence Report," *Bell Journal of Economics*.

[3] Synfuels Interagency Task Force, *Recommendations*, vol. 1, p. 27.

companies an incentive to hold off exploration, development, and production until a future time when controls are lifted.

A more straightforward approach to the problem is to eliminate special price controls on existing conventional fuels. Such action will simultaneously encourage exploration and promote conservation. As the price of conventional fuel rises to the cost of new energy, private companies will have an automatic incentive to move ahead. But given the normal desire to minimize risk, private investments will be deterred as long as there is a strong possibility that the federal government will step in and assume the risk.

To those who are concerned that rising energy prices will be inflationary, it should be pointed out that inflation is not curbed by holding down individual prices (the classic way to create a shortage). The basic way to reduce inflationary pressures is well-known—to deal with the forces that influence the overall price level by reducing the budget deficit and maintaining a moderate monetary policy.

Although it is too early to determine the precise long-term impact of price increases on the use of energy, preliminary indications confirm the direction of the expected change: higher prices of goods or services cause consumers to curtail buying. One regional administrator of the U.S. Federal Energy Administration estimated that in New York and New Jersey residential heating oil was used 15 percent less in the winter of 1975 than in the preceding winter, after adjusting for weather differences over the two winters. In January 1975 the general counsel of the Connecticut Energy Agency was reported in the *New York Times* as stating that people on fixed incomes had of necessity cut back on heating oil and that there was a move to conserve energy in office and commercial buildings.[4]

Government in Business. Other shortcomings of the EIA proposal are evident. If it had developed as envisioned by its sponsors, the Energy Independence Authority would have represented a large involvement of government in sectors of the economy traditionally the responsibility of private enterprise. In the words of former energy administrator John C. Sawhill, "The proposed Energy Independence Authority . . . represents a major new intrusion into the private sector."[5] Through this new agency, the federal government would have become a prominent factor in the financing of energy industries, which it is far from being at present.

[4] Joseph P. Fried, "Energy Savings Here Tied to Higher Costs," *New York Times,* January 12, 1975, p. 1 ff.
[5] John C. Sawhill, "What Makes America Work? Energy . . . and It's Time We Became Independent," *Wall Street Journal,* December 2, 1975, p. 15.

To the extent that the federal government, through the EIA, would have stood ready to share the financing of new projects, private capital would have been less likely to finance the more risky energy undertakings on its own. But EIA would have been more than a financing mechanism. Since the agency would have been authorized to make "high risk" loans, some private borrowers certainly would have defaulted. EIA would have wound up owning the collateral, that is, the projects that it had financed. As pointed out above, the proposed law clearly envisioned circumstances under which EIA would be operating commercial energy projects as a result of con-templated defaults. In fact, the proposed law went on to require that, under such circumstances, EIA would pay taxes in the same manner as private business firms. The cost-benefit analysis prepared by the Synfuels Task Force supports the conclusion that many of these projects would be uneconomical. Hence, their ultimate financial viability would be subject to considerable doubt.

So long as EIA met the rather vaguely worded "restrictions" imposed by the pending legislation, its board of directors would have great discretion in selecting the companies to receive financial aid, the types of financial assistance that it would offer, and the specific terms on which the assistance would be provided. Unfortunately, the history of government credit agencies exercising broad discretion is not comforting.

During the 1930s and 1940s, the Reconstruction Finance Cor-poration, a federal enterprise financed with many billions of dollars of Treasury debt and tax receipts, undertook a wide variety of activities. Without underestimating the contributions of the RFC's early years, it is pertinent to note that Congress ended the agency's existence amid a rash of allegations of improper activities. An ex-tended Senate investigation disclosed gross abuses of the power and authority vested in the enterprise.

It is important to understand that when Congress enacts a credit program, such as the Energy Independence Authority, the total amount of investment funds available to the economy is not increased. The program merely gives one group of private borrowers a preferred position over other private borrowers, a matter of robbing Peter to pay (or lend to) Paul.

The demands for new federal credit programs are, not sur-prisingly, insatiable. Whenever one group is singled out for prefer-ence, another group asks for similar treatment. Who gets squeezed out? New and small businesses, school districts and smaller local governments, and individuals—generally the weaker borrowers. The unsubsidized private borrowers wind up paying higher interest rates.

Budget Subterfuge

The establishment of EIA would have been a major extension of off-budget financing of federal government activities. To a far greater extent than at present, the reported totals of revenues and expenditures would have understated the true magnitude of governmental activities. The reported budget deficits would have become less meaningful measures of federal financing needs. The counter-argument that EIA's credit extensions would have been repaid is not persuasive.

Many programs in the budget do generate offsetting revenues, including the Tennessee Valley Authority, the Commodity Credit Corporation, the Farmers Home Administration, the Federal Crop Insurance Corporation, the Government National Mortgage Association, and the National Credit Union Administration. Clearly, the expectation that a federal spending program will ultimately yield receipts is not a sufficient justification for excluding it from the budget. There are credit programs which properly have been excluded, and their transactions do not appear in the totals of federal revenues and expenditures. But each of these latter organizations—such as the Federal National Mortgage Association, the Federal Land Banks, and the Federal Home Loan Banks—have repaid the Treasury's original investments and are now privately owned. In the case of the Energy Independence Authority, on the other hand, all of its capital stock would have been held by the secretary of the Treasury, and all of its debentures would have been guaranteed by the full faith and credit of the U.S. government.

Administrative Problems. To avoid the scandals that led the Congress to terminate the Reconstruction Finance Corporation, the proposed EIA statute stipulated restrictions on the corporation's lending ability, criteria for the guidance of its directors, and reviews by other federal agencies. It is difficult to see how these vaguely worded and often conflicting statements could truly have produced better results, though the time and expense involved in following the guidelines would have been substantial.

For example, EIA could not have given financial assistance to projects unless the applicant had shown "satisfactory" efficiency, management capacity, and other factors usually taken into consideration by private sources of financing. Yet EIA financing would have been limited to projects unable to obtain adequate financing from private sources, presumably because those firms were not satisfied with the efficiency, management capacity, and so on, of the same project.

The proposed statute expressed a general concern for an economically sound and competitive private sector. Yet, in many ways it would have required the recipients of EIA assistance to conduct themselves as federal agencies and federal contractors do. This is the case in respect to adherence to federal labor standards, including the Davis-Bacon Act, and equal-employment-opportunity requirements. On the one hand, the EIA bill would have given the Federal Energy Authority new power to expedite the regulatory process for energy projects. But, on the other hand, it would have set up a new level of reviews, requiring the Energy Resources Council, among other federal agencies, to examine each project prior to EIA's extending it any financial aid.

Regulatory Barriers to Domestic Energy Production. The expansion of federal regulation has resulted in a new obstacle to carrying out large developmental projects. Virtually every proposed energy project has been delayed by this factor for an unpredictable length of time and at significant cost. The EIA proposal merely set up a new effort to offset in part the adverse effects of earlier federal efforts. The uncertainty produced by enforcement of the various environmental programs is one such adverse effect. The Synfuels Interagency Task Force commented on the Federal Water Pollution Control Act Amendments of 1972: "It would be next to impossible at this time to predict the impact of these requirements on synthetic fuels production."

It is hard to believe that regulatory bodies such as the environmental agencies have taken account of the impact of their activities on other national objectives, such as fostering domestic energy independence.

Summary

The proposed program of credit subsidies via an Energy Independence Authority to promote domestic energy development was undesirable for many reasons:

- It avoided dealing with the fundamental need to provide basic market incentives to increase domestic energy production. Although intended to aid private industry, it would have weakened the risk-bearing and entrepreneurial character of the American business system.

- There is no indication that the program would have resulted in any specific increase in domestic energy production.
- It would have been an extremely cumbersome program to operate, involving many government agencies.
- It ignored the sad lessons of history, notably the RFC experience.
- The federal government might have wound up operating commercial plants and selling the products or energy they produced.

In the formation of public policy toward fostering a greater degree of domestic energy independence, decision makers need to consider the various alternative ways of promoting the nation's objectives in the energy area—including greater reliance on normal market incentives, reducing the severe regulatory barriers to use of existing energy sources and to development of new ones, stockpiling petroleum to reduce the threat of embargo, and encouraging more effective use of existing energy supplies.

Despite the high hopes of EIA supporters, there is no assurance that the many billions of dollars earmarked for the Energy Independence Authority would have resulted in the attainment of energy independence. Perhaps the fundamental failure of the measure's advocates is their inability to demonstrate the superiority of its approach—subsidies to the private sector—to other alternatives.

The most basic and satisfying alternatives to federal credit proposals such as the Energy Independence Authority are measures to deal with the underlying conditions of which they are symptoms. If we can create an economic climate conducive to private saving and investment, the need for private borrowers to seek federal aid will be reduced. Achieving an improved economic climate will require reducing deficit financing, maintaining a moderate monetary posture, and adopting tax reforms that encourage private saving.

Cover and book design: Pat Taylor